Contents

OPPOSITES ATTRACT

Opposites Attract

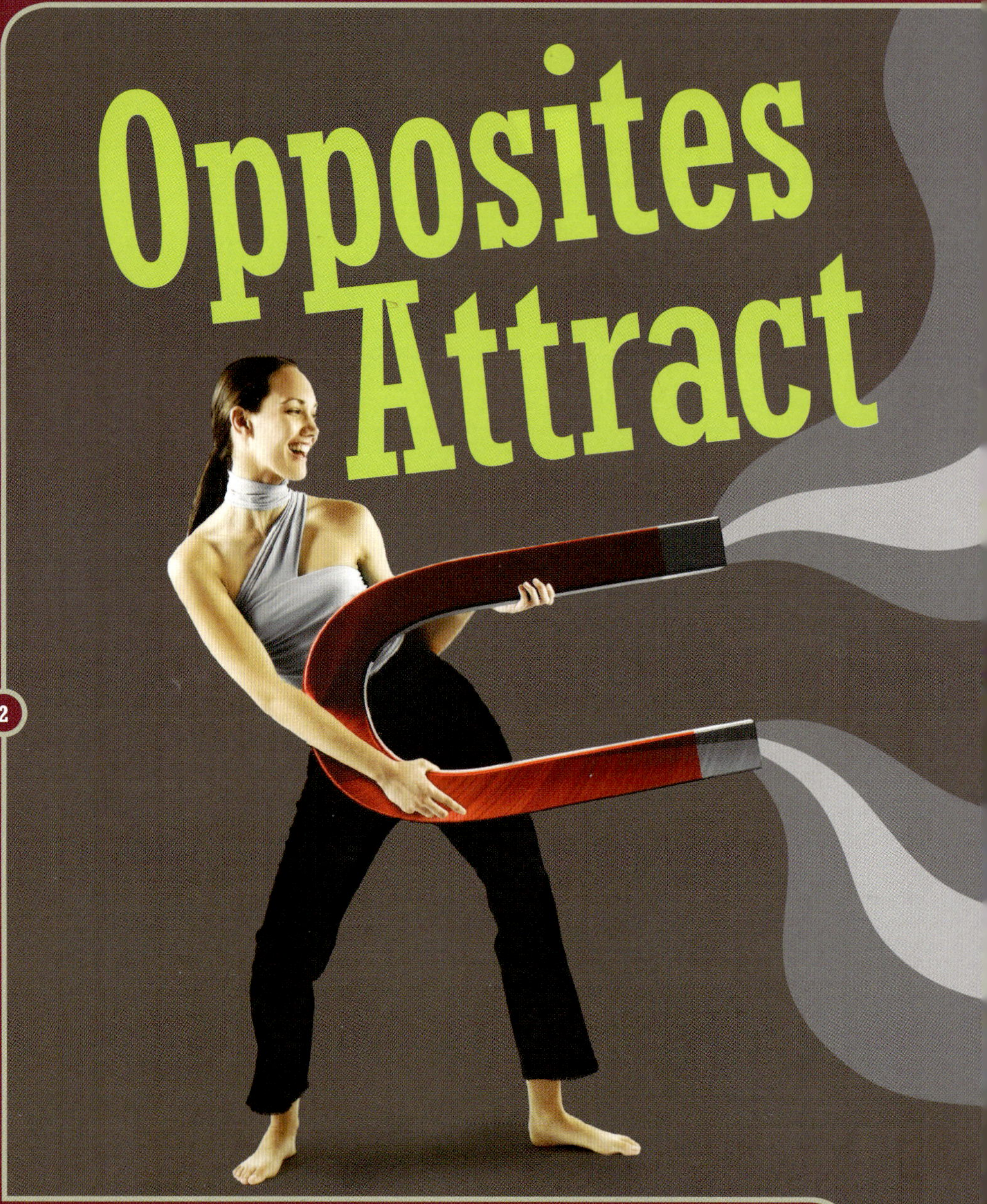

Read these words then check out pages 30 and 31.

attract
uh-TRAKT

electricity
el-ek-TRIS-uh-tee

electromagnet
el-ek-troh-MAG-nit

magnetic field
mag-NET-ik FEELD

Did you know magnets have two poles, like Earth? Check out page 6 to learn more.

Do you know how a compass works? Find out on page 8.

Do you know what the biggest magnet in the world is? Find out on page 11.

Did you know you can make your own magnet? Learn how on page 14.

Did you know that there is a train that runs on magnets? Learn about it on page 25.

magnetic force
mag-NET-ik FAWRS

magnetic poles
mag-NET-ik POHLZ

magnetite
MAG-nuh-tyt

repel
ruh-PEL

Read this article to find out how magnets work. →

what are magnets?

written by collette manners

Magnets are objects with a special power, or **magnetic force**. You cannot see the force. But it is there. It can push, or **repel**, things. It can pull, or **attract**, things. How does it work? Magic? Read on to find out.

Poles

Do you know the North Pole?
It is the place on Earth that is furthest north.
Do you know the South Pole?
It is the place on Earth that is furthest south.
Earth has two other poles as well.
They are not in the same places.
They are the **magnetic poles**.

Magnets have
magnetic poles, too.
They have a north pole.
They have a south pole.
The poles are the ends
of the magnet.

The poles of a magnet are
the same as Earth's poles.
Take a bar magnet.
One end says *N*.
The other says *S*.

Iron filings let you see how the magnetic force works.

The poles are strongly magnetic.
They have strong magnetic force.
They attract strongly.
They repel strongly, too.

Take two magnets.
Put the two north poles together.
They repel each other.
Put the two south poles together.
They repel each other.
Put one north and one south pole together.
They attract each other.

Like poles repel.

Opposite poles attract.

Tie a string around the middle.
The magnet will move.
It will move until the *N* points north
and the *S* points south.
The needle in a compass is a magnet.
It always points north.
It tells you which way is north.

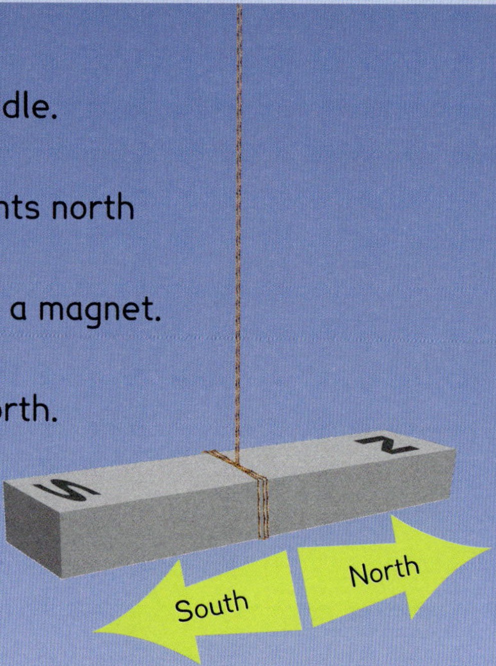

A compass always points north.

magnetic fields

Have you heard of force fields?
Maybe you saw them on TV.
A magnet has a force field.
It is called a **magnetic field.**
The field is the area around the magnet.
It is the area where the special force works.

The field has a special shape.
It makes lines around the magnet.

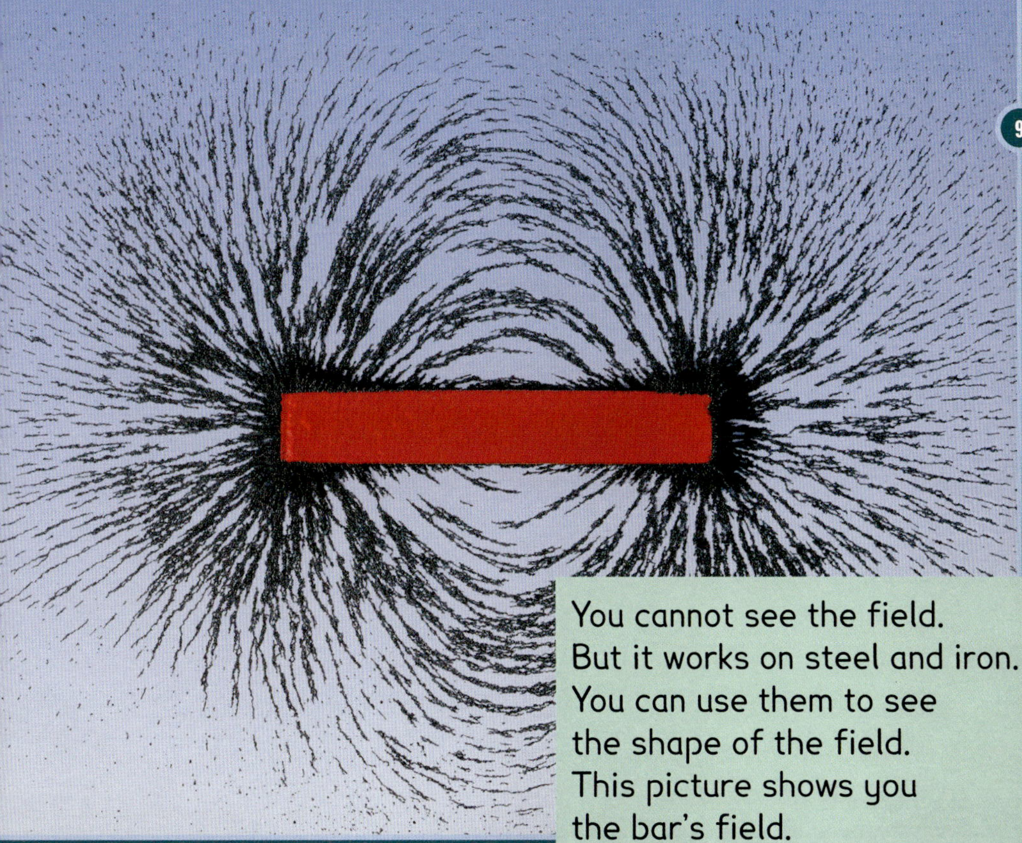

You cannot see the field.
But it works on steel and iron.
You can use them to see
the shape of the field.
This picture shows you
the bar's field.

You can change a magnet's shape.
But the poles do not change.
You can cut a bar magnet.
Now you have two pieces.
Each piece has a north pole.
Each piece has a south pole.

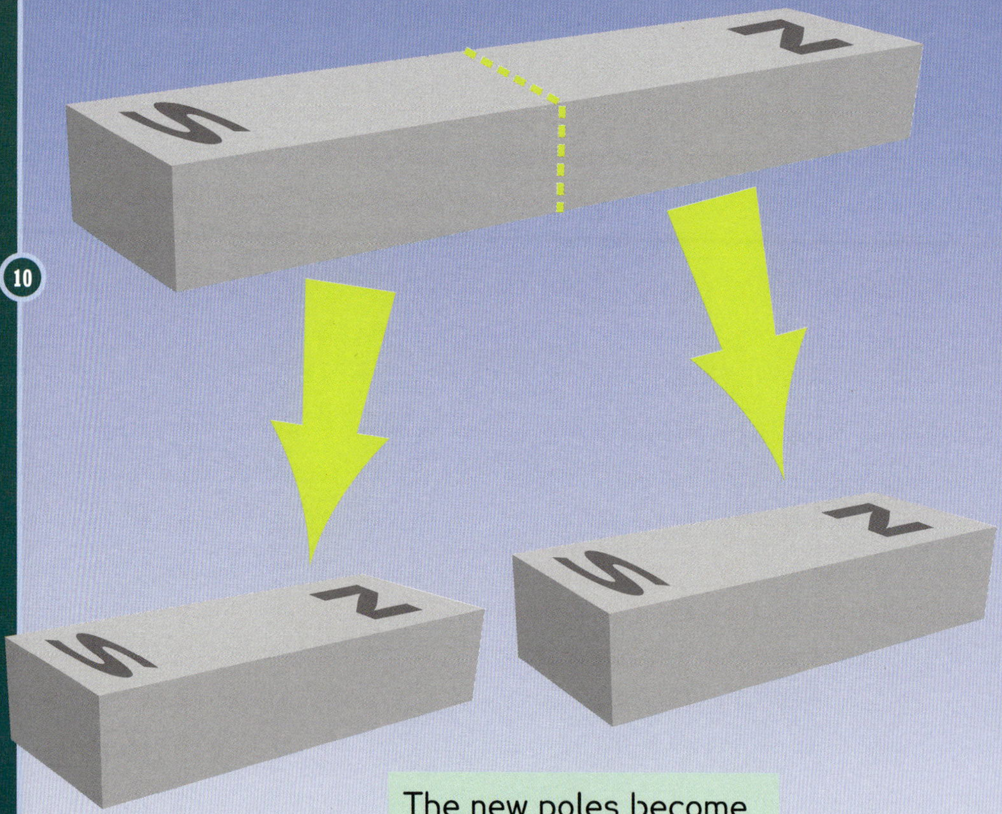

The new poles become strongly magnetic when the magnet is cut.

What is the biggest magnet on Earth?
Can you guess?
The answer is – Earth!
Earth is a giant magnet.
Earth has a magnetic field.
Earth's magnetic field goes into space.
It goes for thousands of kilometres.

The pattern of Earth's field is the same as an ordinary magnet's.

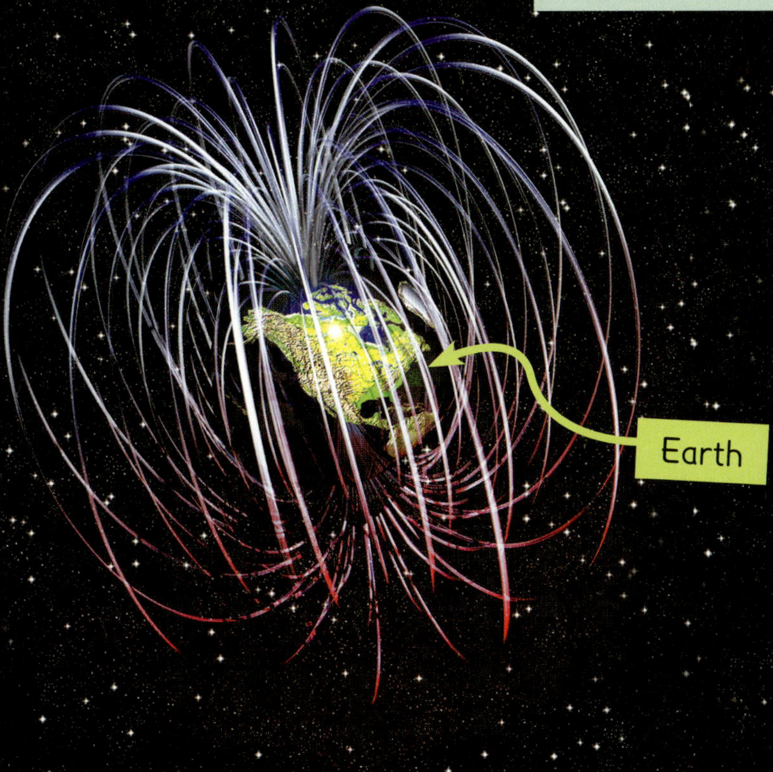

Read on to find out how magnets and electricity are related. →

Magnets and Electricity

Written By Collette Manners

Magnets attract and repel.
Electricity is the power that runs machines.
They are different. Aren't they?
No. They are the same thing.
They both use the special magnetic force.

A magnet can make electricity.
Move a piece of wire near a magnet.
It makes an electric current.

Electricity can make a magnetic field.
Pass electric current through a wire.
The wire now has a field.
It is an **electromagnet**.

Why must you use a wire?
Why not a straw?
Or a piece of chalk?
The wire is made of iron.
Electricity can pass through iron.

Magnets attract wire.

Magnets work on iron too, don't they?
A magnet can pick up a wire.
It can't pick up a straw.
It can't pick up a piece of chalk.

Magnets and electricity work on iron.
Iron is strongly magnetic.
They do not work on plastic.
They do not work on chalk.
Plastic and chalk are weakly magnetic.

Magnets do not attract straws or chalk.

Read on to find out how to make your own magnet. →

How to Make a Simple Electromagnet

Written by
Collette Manners

Make your own simple electromagnet.
You will need –

- copper wire
- a nail
- a D-cell battery
- a battery holder
- paper clips

What to Do –

1. Wrap the wire around the nail.
2. Join the wire to the battery holder.
3. Put the battery into the holder.
4. Hold the nail over the paper clips.
5. See what happens!

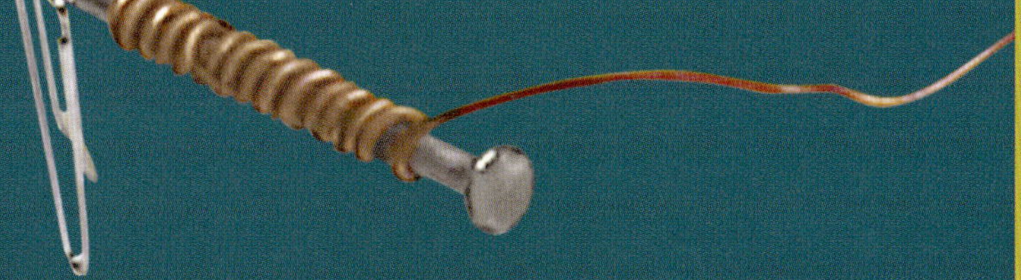

You can switch your magnet on and off. Just take the battery out of the holder.

Read on to find out about magnets in daily life. →

How Do People Use Magnets?

Written by Collette Manners

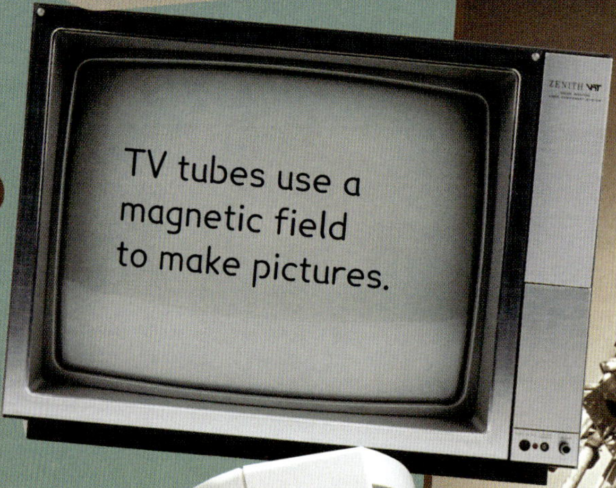

TV tubes use a magnetic field to make pictures.

Magnets pick up rubbish in a wrecker's yard.

Hard disks use magnets to store memory.

A magnet also keeps the fridge door closed.

Compasses help people find their direction.

The stripe on an ATM card is made of tiny magnets.

The super-fast maglev train runs on a magnetic track.

Read on to see the power of magnets! →

THE BIRTHDAY MAGNETS

Written by Collette Manners • Illustrated by Matt Wilson

Juan and Olivia were twins. They got their birthday presents on the same day. They could never agree whose present was better. What do you think?

Hey! A set of magnets! All different kinds!
Magnets? Why do you need magnets?
Next Saturday
Still playing with your magnets?
No, I'm not playing. I'm studying. And you're going camping, are you?
I'm going to the movies.
Still wearing that pack? So you can camp out at the cinema?
Look, Tito! It has a pouch for a GPS watch! I don't even have to take the pack off.
I just reach back like this...

You don't have a GPS watch.
Bye! Enjoy your magnets!
We will, won't we, Tito?
Hmmm. It says compass needles are magnets. One end points south. The other points north. That's how compasses work. They point to the North Pole.
I bet camping boy doesn't know that.
MAGNETS
Uh-oh. Listen to this, Tito. You know the stripe on an ATM card? The stripe is made of magnets.
It has a lot of tiny bar magnets – like those ones, but much smaller.
MAGNETS

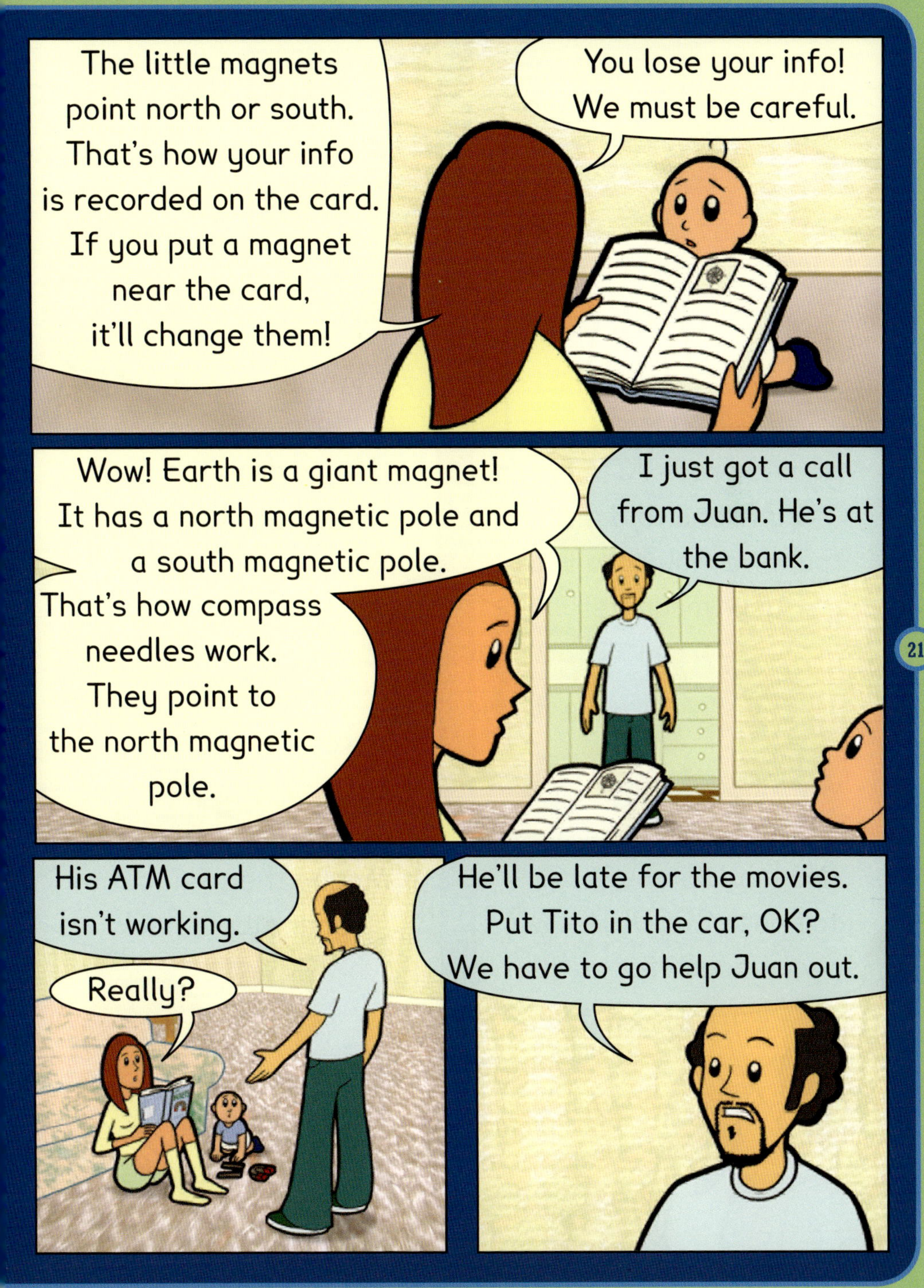

The little magnets point north or south. That's how your info is recorded on the card. If you put a magnet near the card, it'll change them!
You lose your info! We must be careful.
Wow! Earth is a giant magnet! It has a north magnetic pole and a south magnetic pole.
That's how compass needles work. They point to the north magnetic pole.
I just got a call from Juan. He's at the bank.
His ATM card isn't working.
Really?
He'll be late for the movies. Put Tito in the car, OK? We have to go help Juan out.

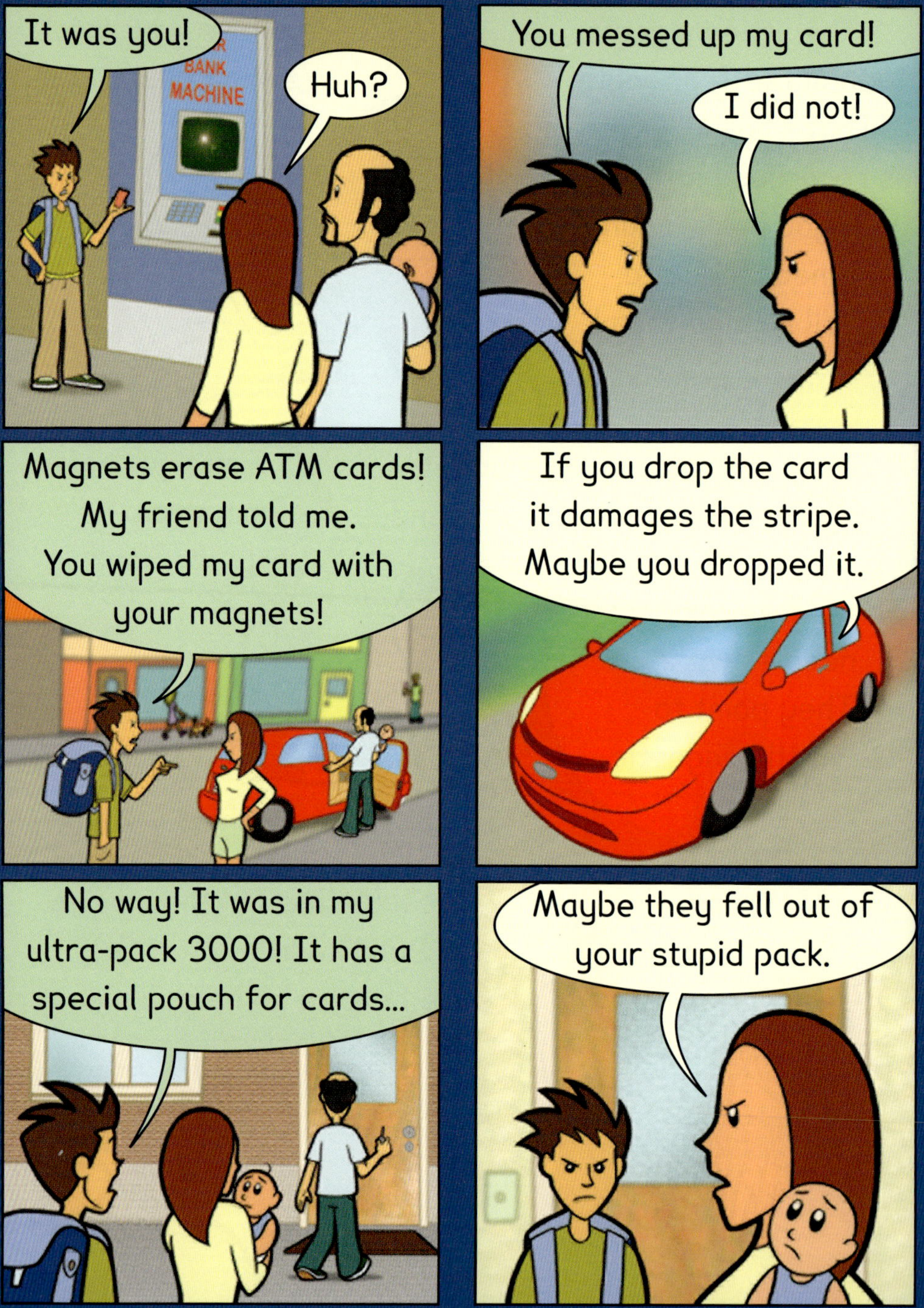
It was you!
BANK MACHINE
Huh?
You messed up my card!
I did not!
Magnets erase ATM cards! My friend told me. You wiped my card with your magnets!
If you drop the card it damages the stripe. Maybe you dropped it.
No way! It was in my ultra-pack 3000! It has a special pouch for cards...
Maybe they fell out of your stupid pack.

Did you know that sand can be magnetic? Turn the page to learn more. →

Multimedia Information

www.readingwinners.com.au

FAQS

Q What is a lodestone?

A A lodestone is a piece of **magnetite**.
Magnetite is a mineral in black sand.
It has two kinds of iron.
Electricity moves between the two kinds.
It makes a magnetic field.

You can rub a needle on a lodestone.
The needle becomes a magnet.
The Chinese used lodestones.
They used them to make the first compasses.

Floating with Magnets

Did you know magnets
can make things float?

There is a special train
that floats on magnets.
It is called a maglev train.
It does not touch the rails.
It can move very fast.

Other things can float
on magnets, too.
Look at the picture.
Can you believe it?

A metal disc
floating in the air

Read on to learn about moving magnetic poles! →

Homework Message Board

Written by John Suzuki

www.readingwinners.com.au

From: Calvin "Need Help with Homework" Cable

I've heard the magnetic poles can move. True?

From: Gilda "Famous Scientist" Gable

Re: I've heard the magnetic poles can move. True?

True. The North and South Poles do not move.
But the magnetic poles move.
They move all the time.

The magnetic north pole is in Canada.
But it is moving.
It moves fast.
It moves 40 kilometres every year.
It is moving towards Russia.

www.readingwinners.com.au

From: Gilda "Famous Scientist" Gable

Sometimes the magnetic poles switch.
North is now south!
South is now north!
This has happened many times.
The last time was almost a million years ago.

Before it happens, Earth's field weakens.
It is weakening now.
It was stronger just 150 years ago.
The poles may switch again.
What will happen then?
No one knows.

From: Calvin "Need Help with Homework" Cable

Thanks, Prof. That's a great help.

Turn the page to check what you have learned. →

8 Quick Quiz

1. What happens if you put two south poles together?
2. How can you see a magnetic field?
3. What happens if you cut a magnet in half?
4. What is the biggest magnet on Earth?
5. How do people use magnets to shop?
6. How do people use them to travel?
7. Who made the first compasses?
8. In which country is the north magnetic pole?

Turn to page 32 for clues. →

Learn More

Choose Your Topic

Choose one way people use magnets in this book.
It may be a magnet that is used to pick things up.
It may be a machine that has a magnet inside.

Research Your Topic

Find out where your magnet is used.
Find out what it is used for.
Find out how the magnet fits into the machine.
Find out how the machine works.

Write Your Article

You may need to make notes first.
You may need to find photos.
You may need to draw diagrams.
Get your facts in order.
Use subheadings to help you do this.
Write a draft.
Check your spelling.
Check your pronunciation.

Present Your Topic

Share your work with other members of your group.

attract – to pull things together

electricity – the power that runs machines

electromagnet – a magnet made by passing an electric current through metal

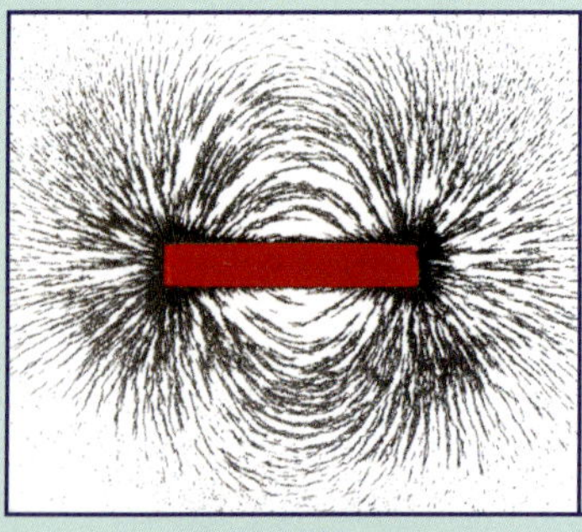

magnetic field – the space around the magnet where the force is

magnetic force – the invisible pushing and pulling power of magnets

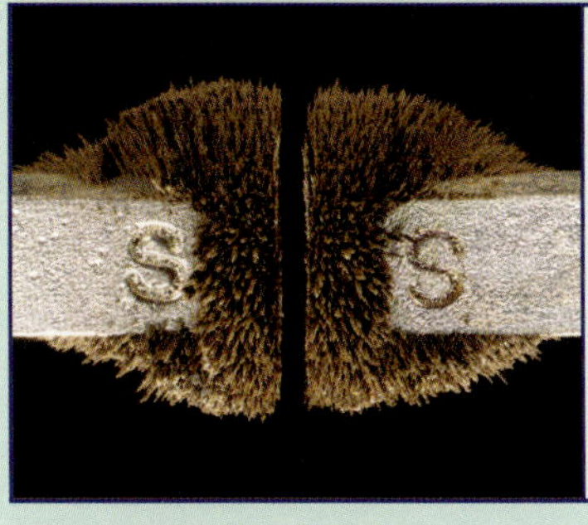

magnetic poles – the ends of a magnet, where the force is strong

magnetite – a magnetic mineral in black sand

repel – to push things away

Index

Clues to the Quick 8 Quiz

1. Go to page 7.
2. Go to page 9.
3. Go to page 10.
4. Go to page 11.
5. Go to page 17.
6. Go to page 17.
7. Go to page 24.
8. Go to page 26.